# DEADLY

# Factbook: Insects and Spiders

Look out for the other
DEADLY books:
OUT NOW :
Deadly Activity Book
Deadly Doodle Book
Deadly Factbook Mammals
Deadly Doodle Book 2
Deadly Annual 2013

BBC EARTH

# DEADLY 60

>>>>> Factbook: Insects and Spiders >>>>>>>>

Orion Children's Books

First published in Great Britain in 2012
by Orion Children's Books
a division of the Orion Publishing Group Ltd
Orion House
5 Upper St Martin's Lane
London WC2H 9EA
An Hachette UK Company

1 3 5 7 9 10 8 6 4 2


Derived from and based on the BBC TV series Deadly 60.

Photo credits
1 © BBC 2009; 6 Pascal Goetgheluck/Ardea; 7 John Mason/Ardea; 11 Nick Gordon/Ardea; 12 © BBC 2009; 13 © BBC 2009; 14 Don Hadden/Ardea; 16 John Clegg/Ardea; 20 Geoff du Feu/ Ardea; 21 © BBC 2009. 22-23 © BBC 2009; 24 John Cancalosi/Ardea; 25 © BBC 2010; 28 Alexandr Pakhnyushchyy/istockphoto; 31 © BBC 2009; 33 © BBC 2010; 34 Steve Hopkin/ Ardea; 36 ifong/Shutterstock; 38 © BBC 2009; 39 Anthony Boulton/istockphoto; 42 © BBC 2010; 43 © BBC 2010; 44 © BBC 2009; 45 Steffen & Alexandra Sailer/Ardea; 46 Andrey Zvoznikov/Ardea; 47 Auscape/Ardea; 48 Auscape/Ardea; 49 © BBC 2010; 50 © BBC 2009; 52 © BBC 2010; 56 James Carmichael Jr/NHPA; 57 Hans D Dossenbach/Ardea; 60 John Cancalosi/Ardea; 62-63 David Chapman/Ardea; 65 Jim Zipp/Ardea; 66 Vinicius Tupinamba/Shutterstock; 68-69 © BBC 2009; 70 John Clegg/Ardea; 74 James Steidl/Shutterstock; 76 © BBC 2010.

Compliled by Jinny Johnson Designed by Sue Michniewicz



A catalogue record for this book is available from the British Library.

ISBN 978 1 4440 0635 3

Printed and bound in Italy by Printer Trento

www.orionbooks.co.uk

# CONTENTS

# WHAT IS AN INSECT?

Insects are incredibly successful animals and there are more kinds of insect on Earth than any other type of creature. As many as three-quarters of animal species may be insects.

An insect's body is divided into 3 parts: head, thorax and abdomen. It has 6 legs and, at least at some stage of its life, usually 2 pairs of wings. Most insects also have 2 antennae on their head.

ELEPHANT HAWK MOTH

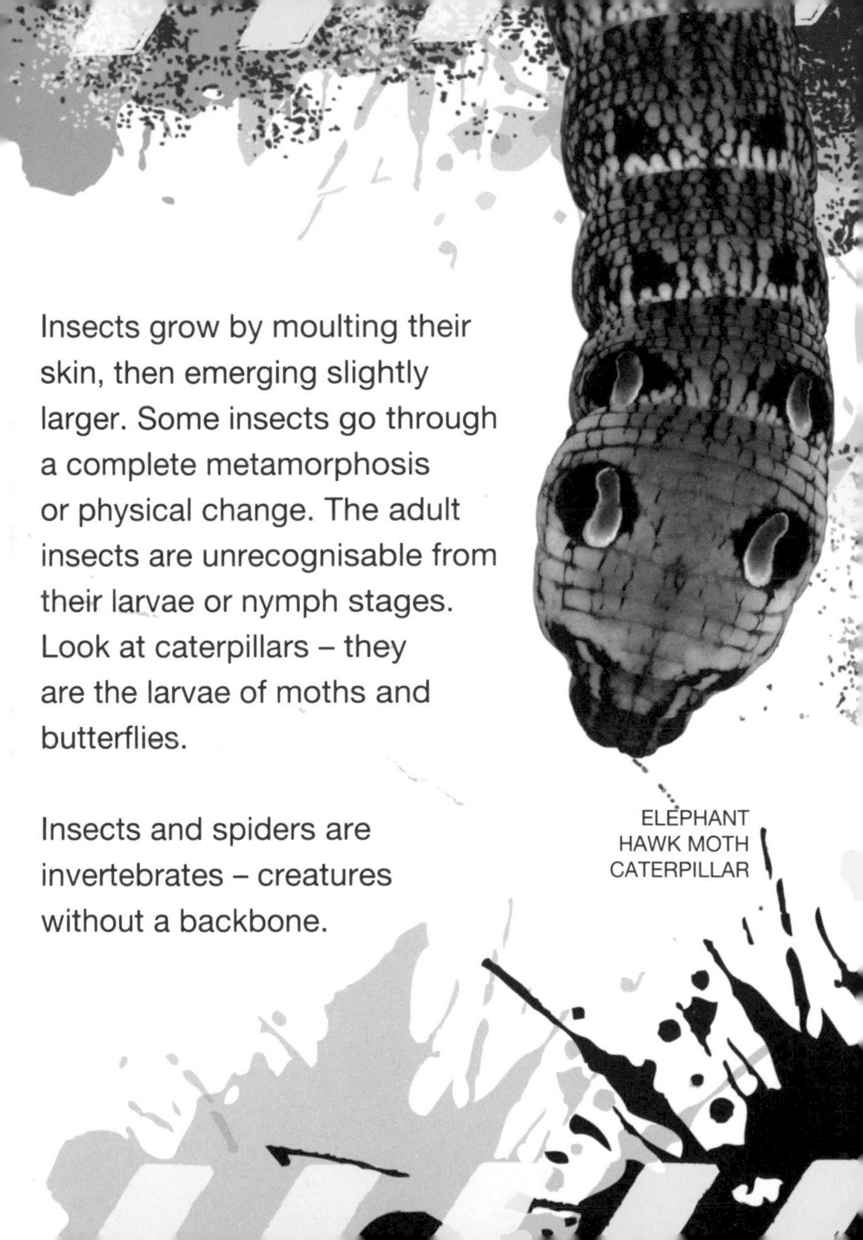

Insects grow by moulting their skin, then emerging slightly larger. Some insects go through a complete metamorphosis or physical change. The adult insects are unrecognisable from their larvae or nymph stages. Look at caterpillars – they are the larvae of moths and butterflies.

Insects and spiders are invertebrates – creatures without a backbone.

ELEPHANT HAWK MOTH CATERPILLAR

# WHAT IS A SPIDER?

A spider is not an insect. It belongs to a separate group called arachnids, which also includes scorpions, ticks, whip spiders and solifugids.

There are more than 40,000 known species of spider and nearly all are predators.

A spider has 8 legs and a body that is divided into 2 parts. It does not have wings or antennae, but it does have fang-bearing jaws called chelicerae.

# BIGGEST and SMALLEST

## Chapter 1

The **TITAN BEETLE** is an insect giant. This magnificent monster lives in the jungle in South America and the largest found so far measured 16.7 centimetres long. That's as big as the span of an adult human hand!

This beetle isn't a predator but it is an imposing sight. It frightens off any enemies by making loud hissing sounds with the breathing holes along the sides of its body, which are called spiracles. It has strong jaws and male titans have sharp spines on the insides of their legs.

Beetles are the largest group of animals. There are at the very least 350,000 species, more than any other type of animal, and new species are still being discovered.

The huge **GOLIATH BIRD-EATING SPIDER** is the heaviest of all spiders. This truly impressive creature, which lives in South American jungles, weighs up to 175 grams – more than a large apple. True to its name, it can catch birds but more often preys on insects, such as crickets, cockroaches and beetles, as well as rats, mice and lizards.

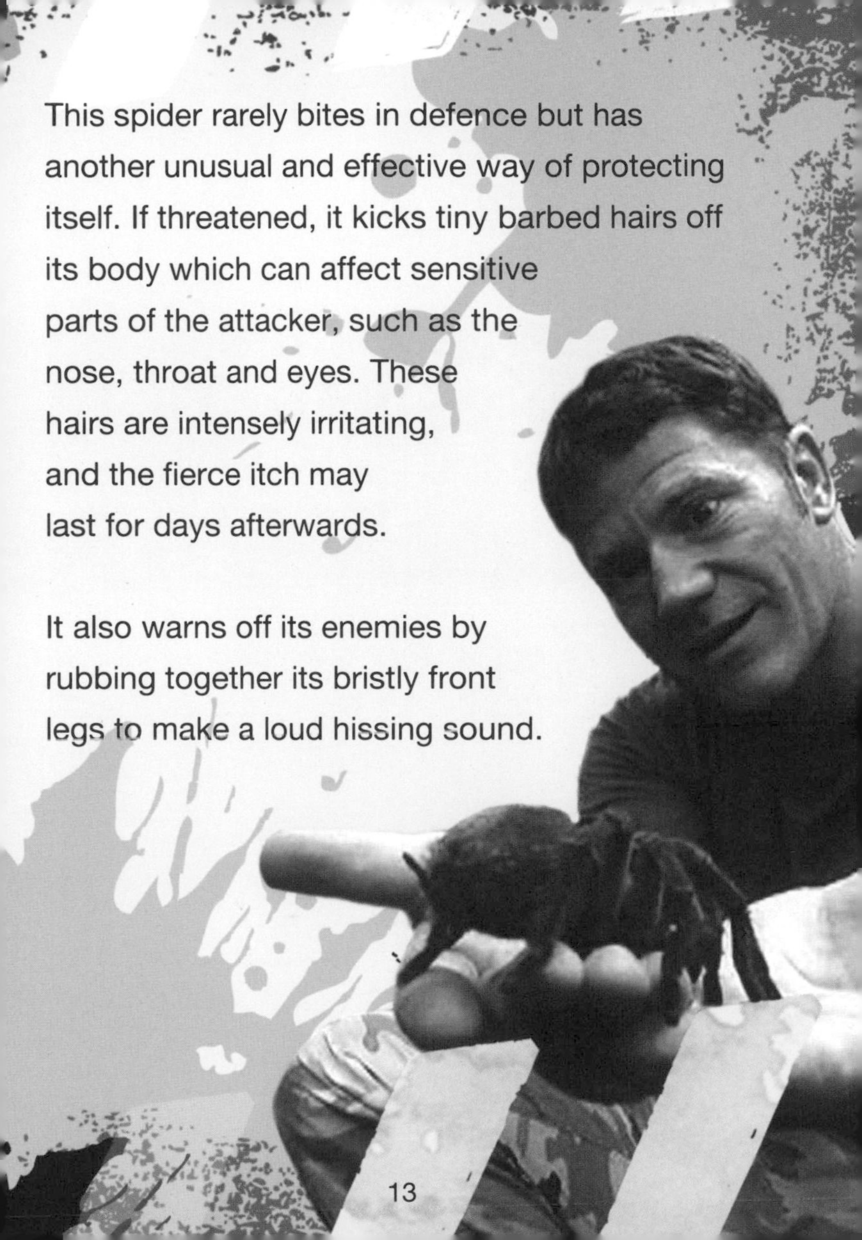

This spider rarely bites in defence but has another unusual and effective way of protecting itself. If threatened, it kicks tiny barbed hairs off its body which can affect sensitive parts of the attacker, such as the nose, throat and eyes. These hairs are intensely irritating, and the fierce itch may last for days afterwards.

It also warns off its enemies by rubbing together its bristly front legs to make a loud hissing sound.

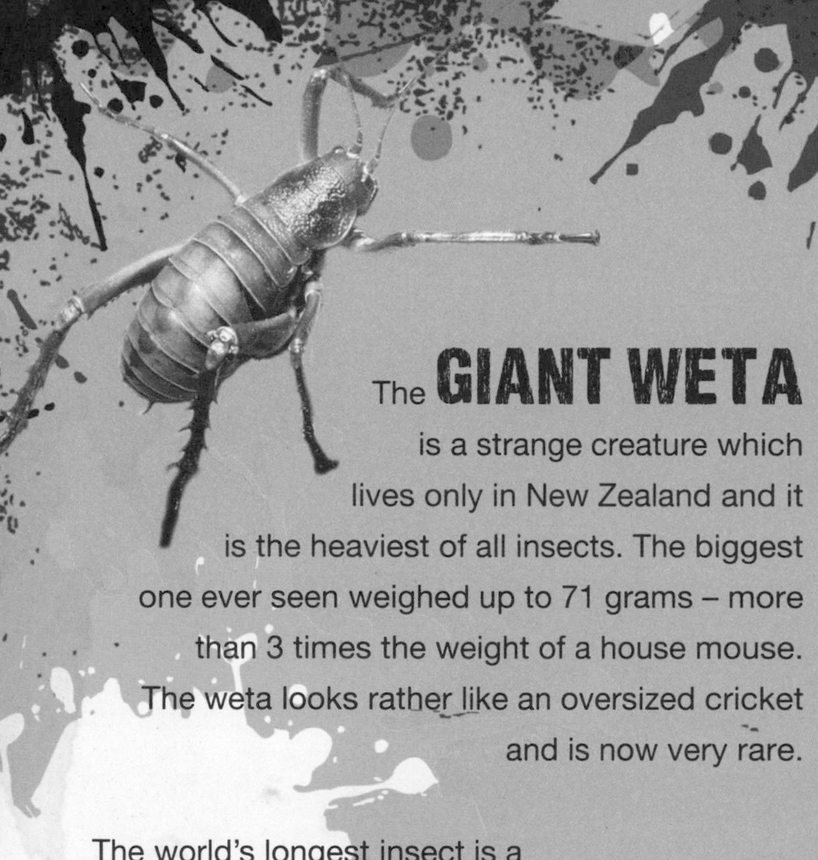

The **GIANT WETA** is a strange creature which lives only in New Zealand and it is the heaviest of all insects. The biggest one ever seen weighed up to 71 grams – more than 3 times the weight of a house mouse. The weta looks rather like an oversized cricket and is now very rare.

The world's longest insect is a **STICK INSECT** found in the forests of Borneo in Southeast Asia. It measures an astonishing 56.7 centimetres – possibly as long as your arm!

Another giant is the

**BABOON SPIDER** of Africa.

Just the body of a baboon spider measures up to 9 centimetres and it has long hairy legs!

Despite this spider's size its venom glands are small, so its bite isn't that harmful. But its fangs are about the same size and shape as the claws of your pet cat, and they are perfect tools for tearing apart other invertebrates!

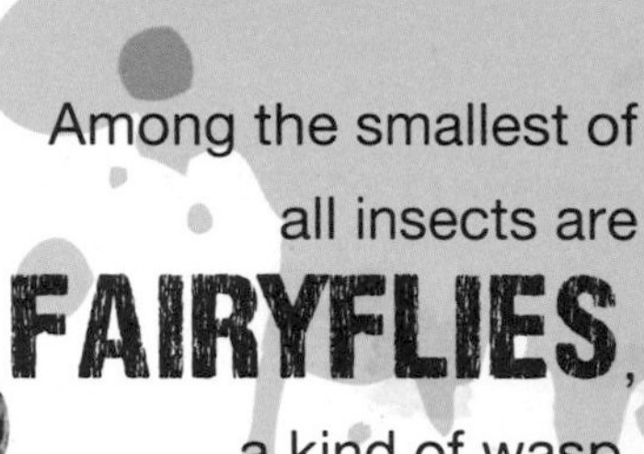

Among the smallest of all insects are **FAIRYFLIES**, a kind of wasp. Some of these measure less than half a millimetre long – way smaller than a pinhead.

These tiny creatures have a great way of making sure their young have food when they hatch. They lay their eggs inside the eggs of other insects. When the young wasps hatch, they will feed on the other baby insects.

THESE PICTURES OF FAIRY FLIES ARE MUCH, MUCH BIGGER THAN LIFE SIZE.

# MICRO HUNTERS

## Chapter 2

# The PRAYING MANTIS

is one of the most ferocious hunters in the insect world. It can turn its head nearly 180 degrees and with its huge compound eyes, composed of thousands of tiny lenses, almost nothing escapes its gaze.

When a meal is within reach, the mantis snaps out its jointed front legs at lightning speed, grabs its prey and has it back in its mouth in a fraction of a second.

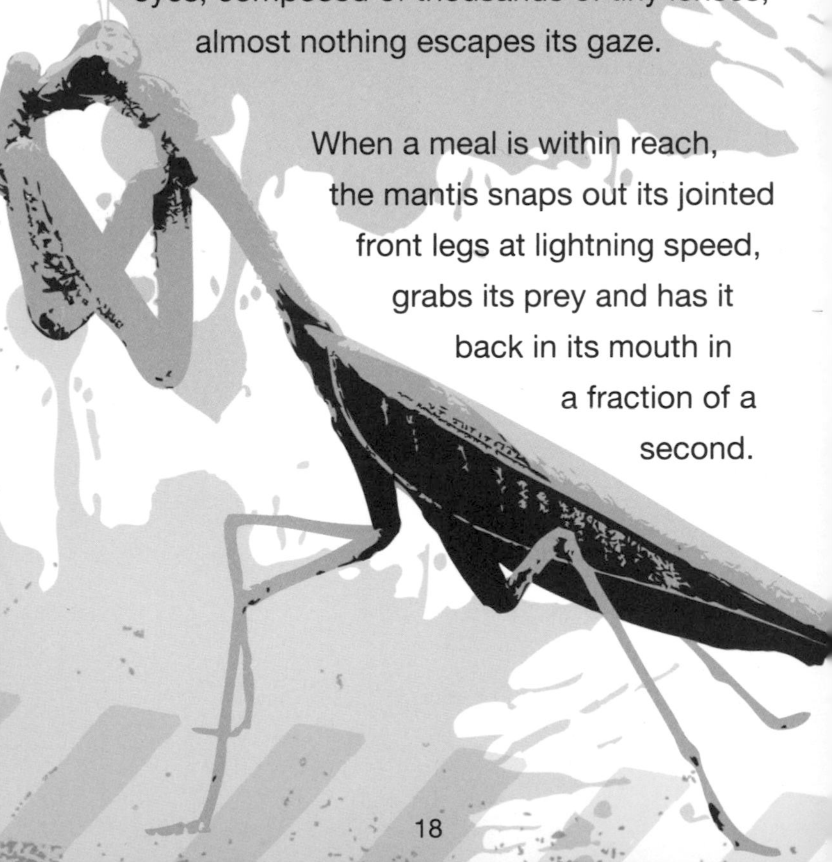

The legs of the mantis are lined with fearsome spikes to help lock on to the struggling catch.

It prefers soft-bodied insects, such as butterflies, flies, moths and grasshoppers, but will also eat its own kind. The female sometimes snaps up the male just after mating!

This fierce little creature gets its common name from its habit of sitting with its front legs clasped together. This makes it look as if it is at prayer.

There are at least 1,800 different species of praying mantis and most live in tropical parts of the world.

Super-sized wasps, **HORNETS** live in colonies – large groups ruled by a female, the queen. Hornets are expert hunters but they don't eat the prey they catch. They take it back to the nest and chew it up to feed to young in the colony. Adult hornets live mostly on nectar and sap.

Although hornets do have a venomous sting they don't usually sting prey – they kill it with their powerful jaws. They use their stings to defend themselves and their nest against attackers.

**DIVING BEETLES** are expert underwater predators. They swim down to the bottom of the pond or river where they lie in wait for prey. When something comes close, the beetle strikes, trapping its victims between its front legs and biting with its strong mandibles – insect mouthparts.

## THE GREAT DIVING BEETLE

is up to 3.5 centimetres long and hunts prey as large as tadpoles and fish. When it dives, it stores air beneath its wing cases. This air is taken through the spiracles (breathing holes) along the sides of the body.

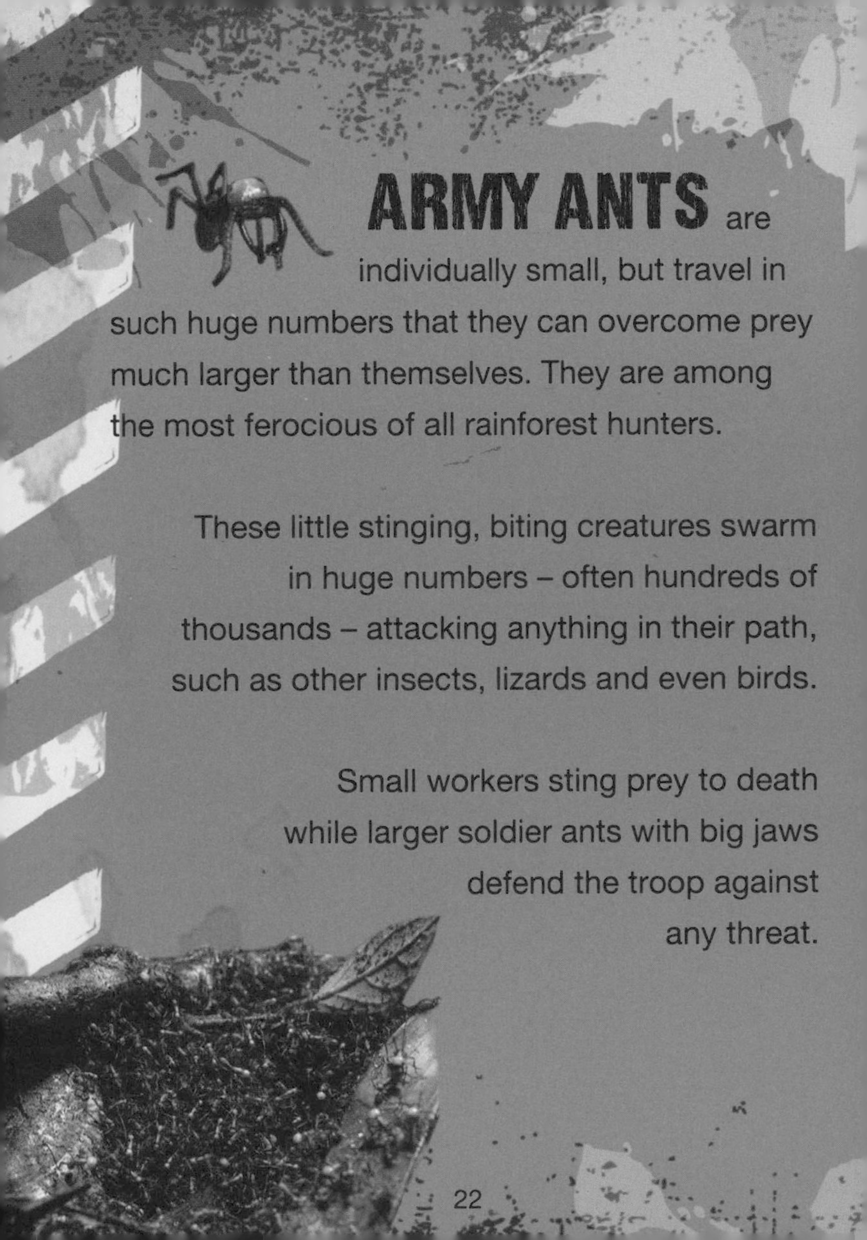

**ARMY ANTS** are individually small, but travel in such huge numbers that they can overcome prey much larger than themselves. They are among the most ferocious of all rainforest hunters.

These little stinging, biting creatures swarm in huge numbers – often hundreds of thousands – attacking anything in their path, such as other insects, lizards and even birds.

Small workers sting prey to death while larger soldier ants with big jaws defend the troop against any threat.

Males and the queen have functioning eyes, but the female worker ants are blind and communicate by smell, sound and touch.

As the ants march, they lay down scent trails for others in the troop to follow. When they stop, the ants make a living bivouac with their own bodies to protect the queen and their young. They carry their young with them as they move from place to place.

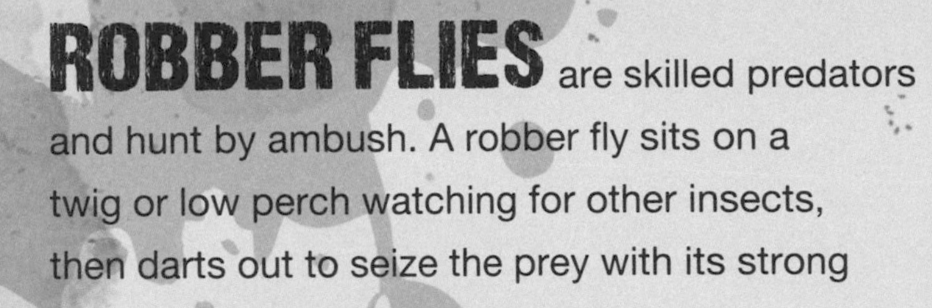

**ROBBER FLIES** are skilled predators and hunt by ambush. A robber fly sits on a twig or low perch watching for other insects, then darts out to seize the prey with its strong stabbing mouthparts.

The robber fly then injects its victim with enzymes – a mixture of paralysing biological compounds that turn the insides to liquid. The fly can then suck the prey's body dry.

## The WHIP SPIDER

is the ultimate cave predator. It is perfectly adapted to hunting in the dark, using its very long front legs for sensing its surroundings. With these, the spider can build up a picture of what is around it.

When it gets close to prey, the whip spider snatches the victim with the barbed claws at the front of its head and draws it into its crushing mouthparts.

This monster is also known as the tail-less whip scorpion. In fact, it isn't a spider or a scorpion but it does belong to the arachnid group.

There may be as many
as 2,000 species of scorpion.
One of the most dangerous is the
**DEATH-STALKER SCORPION**, which
lives in Africa and the Middle East.
It uses its highly toxic venom to kill
insects and other small animals.
It also uses its sting to defend itself.

One of the few scorpions that is dangerous to humans is the

## FAT-TAILED SCORPION.

As well as injecting venom with its sting, it can flick venom into the eyes of its enemies with great accuracy.

The fat-tail is quite large for a highly venomous scorpion, measuring up to 15 centimetres long.

SCORPION PHOTOGRAPHED UNDER ULTRA-VIOLET LIGHT

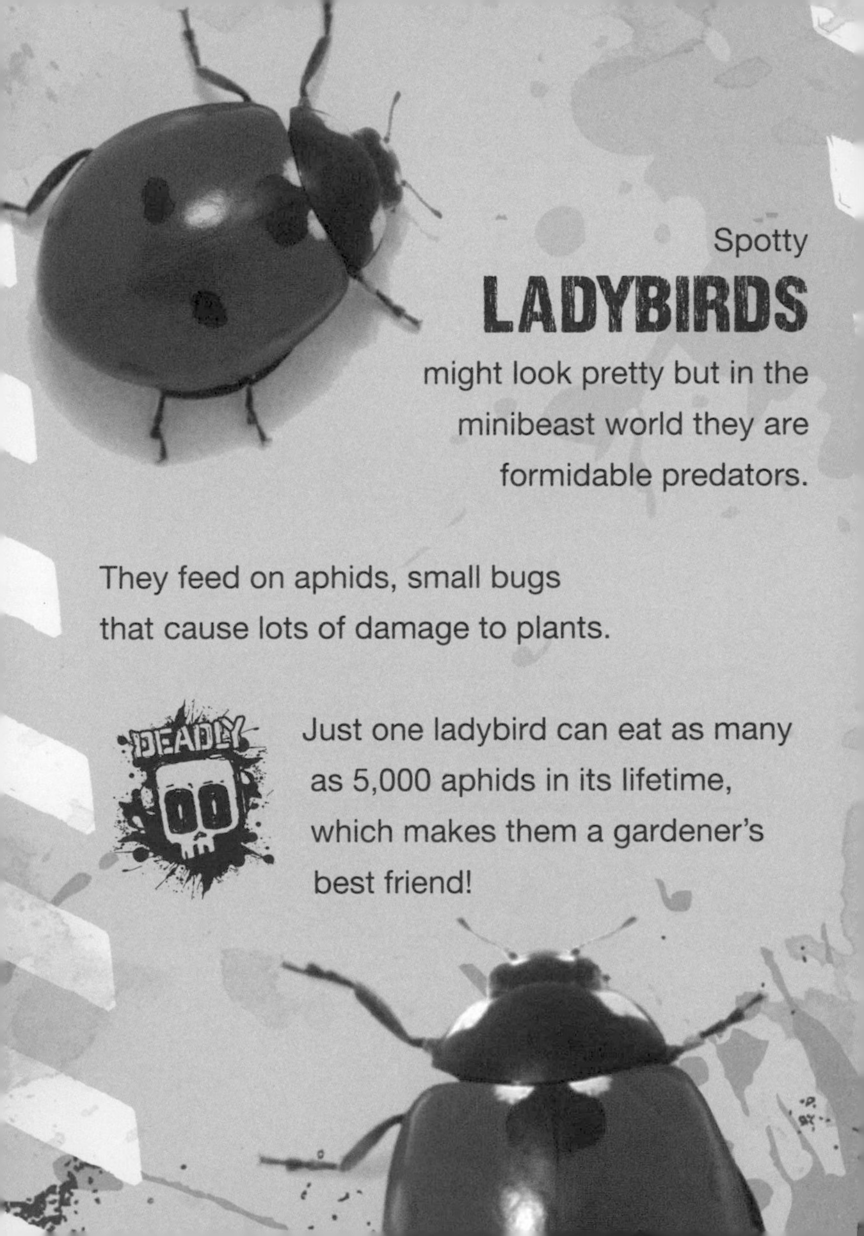

Spotty

# LADYBIRDS

might look pretty but in the minibeast world they are formidable predators.

They feed on aphids, small bugs that cause lots of damage to plants.



Just one ladybird can eat as many as 5,000 aphids in its lifetime, which makes them a gardener's best friend!

# SPEED FREAKS

## Chapter 3

**DRAGONFLIES** are astonishing aerial acrobats. They zoom to and fro as they chase and capture prey in the air and they can fly at speeds of up to 38 kilometres an hour.

Their wings move independently from each other and a dragonfly can fly backwards, forwards and can even hover.

A dragonfly's large eyes are made up of thousands of different units. They are particularly sensitive to movement so ideal for spotting other flying insects.

The dragonfly snatches its prey with its spiny front legs and devours it with its sharp biting mouthparts.

In fact, the scientific name for the group, Odonata, means 'toothed ones'.

Young dragonflies, called nymphs or larvae, lead very different lives from their parents. They live in water for as long as 2 years until they are ready to transform into winged adults.

A nymph hunts other insects, small fish and tadpoles. When a likely meal comes near, it shoots out its extendable, hinged lower jaw, snatching the prey and dragging it back to its mouth.

A dragonfly nymph generally strolls along the river bed, but can also move fast over short distances, propelling itself along by shooting jets of water out of its bottom!

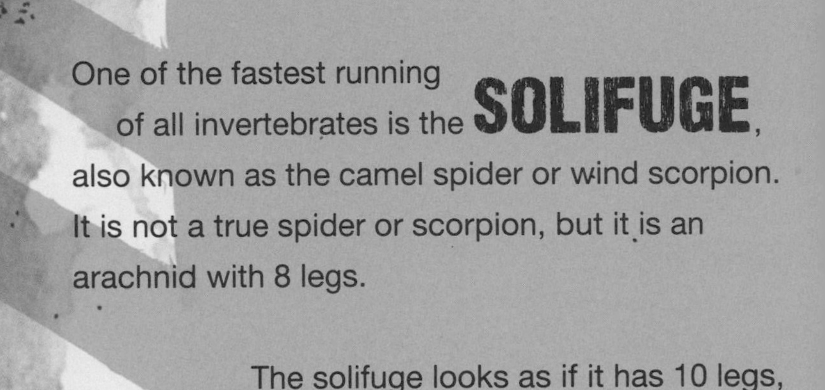

One of the fastest running of all invertebrates is the **SOLIFUGE**, also known as the camel spider or wind scorpion. It is not a true spider or scorpion, but it is an arachnid with 8 legs.

The solifuge looks as if it has 10 legs, but the 2 'front' legs are special sensory organs, used to track down prey.

The solifuge chases its prey at incredible speed and attacks with its oversized jaws.

In fact, proportionally for its size, this awesome arachnid has the biggest jaws of any living animal!

An Australian **TIGER BEETLE** is probably the world's fastest running insect. A fierce hunter, this beetle can reach speeds of 9 kilometres an hour when chasing prey.
That's very fast
for a little insect!

Some **MIDGES** (which are actually tiny bloodsucking flies) beat their wings faster than any other creature. One kind can accomplish an astonishing 1,000 beats a second.

# BIZARRE BODIES

## Chapter 4

Did you know that the **MEXICAN REDKNEE TARANTULA** has a special area on each leg that helps it find prey? These areas are sensitive to smells, tastes and vibrations so help the tarantula track down its food.

Once the spider finds prey it injects it with paralysing venom before tucking in.

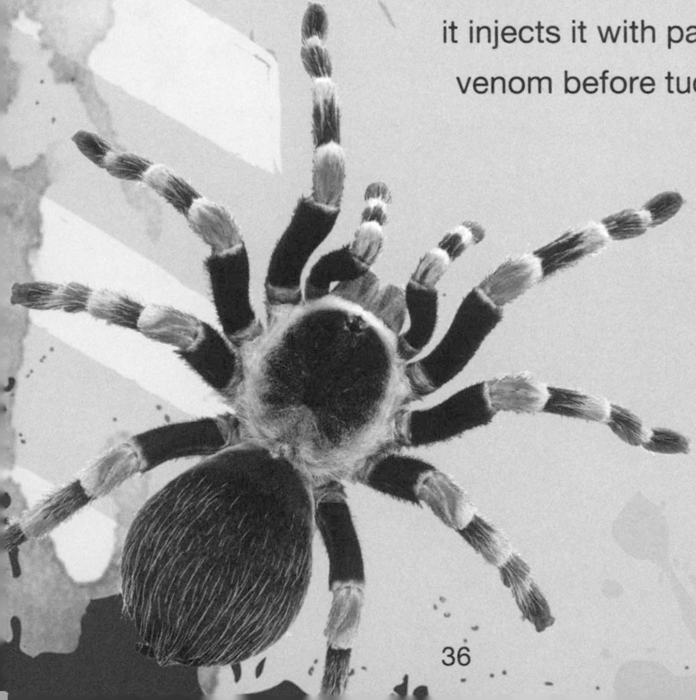

# HONEYPOT ANTS

use some of their workers as living storage jars!

These ants live in very dry areas and food can be scarce at times. They do prey on other insects but also gather flower nectar which they feed to special workers called repletes.

The repletes are fed so much that their abdomens swell up and they can scarcely move. They hang in the nest, ready to regurgitate food for the rest of the colony when needed.

Even though a worker honeypot ant is normally only a couple of millimetres long, when fully swollen with food it can be the size of a cherry. These ants are particularly prized by local Aboriginal people, who consider them a delicacy, and one of the best sources of natural sugars.

Have you ever wondered if insects have ears? They do, but they are not a bit like mammal ears.

In mammals, like us, ears are always on the head, but in insects the hearing organs can be anywhere on the body.

Crickets have 'ears' on their legs, some hawk moths have ears on their mouthparts and mantid ears are between their back legs.

Most adult insects have
2 pairs of wings but **FLIES**
have only 1 pair.
The other pair has become
organs called halteres.
These little club-shaped structures
are believed to help the
fly balance as it flies.

There are more
than 124,000
species of fly
living all over the world.

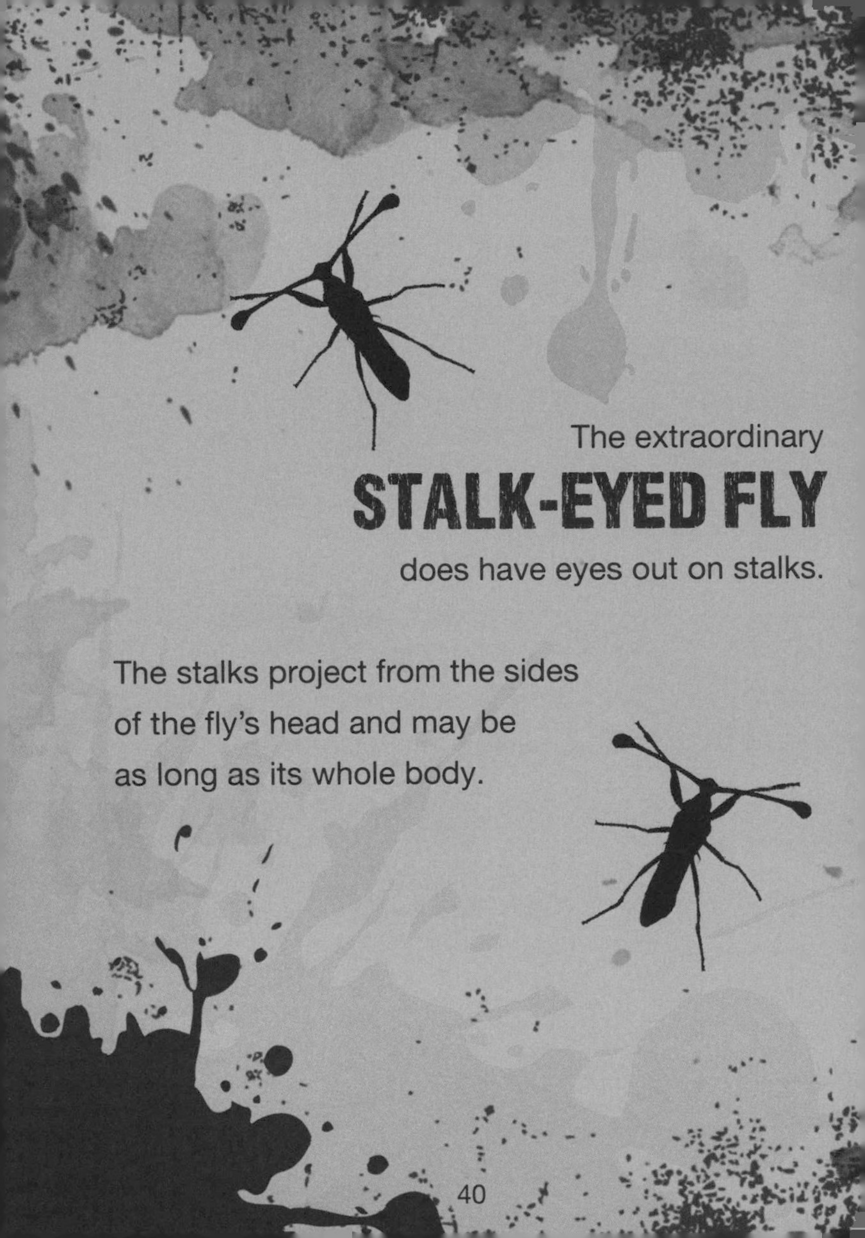

The extraordinary

# STALK-EYED FLY

does have eyes out on stalks.

The stalks project from the sides of the fly's head and may be as long as its whole body.

# TRAPS AND TRICKS

## Chapter 5

The female

# ORB WEAVER SPIDER

builds the best of all insect traps, using silk that she concocts inside her own body. Male orb weavers don't usually make webs.

The spider starts by making a framework of silken threads attached to plants or other structures and then adds a spiral of sticky silk that will trap her victims. When the web is completed, she waits in a position where she can sense any commotion in her web. When an insect flies into the web it struggles desperately to get free.

The movement alerts the spider and the more the insect struggles, the more tangled it gets. The spider comes over, injects the catch with paralysing venom, then wraps it up in silk to make a neat packed lunch.

DEADLY

Weight for weight, silk is stronger than steel.

The female

# GOLDEN ORB-WEB SPIDER

spins the largest of all orb weaver spider webs and is believed to make the strongest silk. The web can be 1-2 metres across and is built to catch flying insects such as flies, wasps and butterflies.

Female golden orb-web spiders can measure up to 50 millimetres long, but the little males are only 6 millimetres.

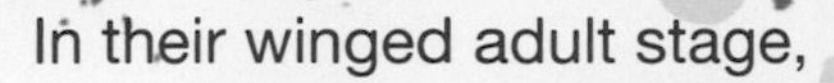

In their winged adult stage,

# ANT-LIONS

look similar to dragonflies, but the larvae are drastically different. They are fierce little hunters, with a special way of trapping prey.

The ant-lion digs a conical pit in sandy ground and sits half-buried at the bottom, waiting to seize other insects that wander by and tumble into the trap. Some ant-lions even flick sand up towards insects on the rim of their pit, knocking them off balance so they fall into the trap. With jaws that can be much bigger than its head, the ant-lion makes short work of its meal.

ANT LION LARVA

Not all spiders spin circular wheel-like webs. The

**TRAPDOOR SPIDER**

lives in an underground burrow with a hinged lid made of silk and earth. The spider surrounds the entrance to the burrow with lines of silk, like trip wires, and then settles down to wait just inside the trapdoor.

When an insect blunders into one of the silken lines, the spider springs out, seizes its prey and drags it inside to eat at its leisure.

# The BOLAS SPIDER

has a special hunting weapon – a line of silk with a sticky ball at the end which it dangles from its front legs. When the spider spots a target such as a moth, it swings its weapon and, with luck, catches the insect on the sticky ball.

This spider has yet another trick. When hunting, it releases a smelly chemical called a pheromone similar to those released by female moths. Male moths are tricked into coming and investigating, and fly straight into the waiting spider's trap.

Beautiful

# CRAB SPIDERS

don't make webs. They are camouflage champions and rely on staying hidden while hunting. Often coloured bright yellow or white, the spider sits on a flower of the same colour, watching for prey, and is very hard to see.

When an unsuspecting insect lands on its flower to feed on nectar, the spider pounces and injects it with deadly venom with a bite from its fangs.

Some of these crafty killers can even change colour to match the flower they are sitting on.

Female **REDBACK SPIDERS** make tangled webs that are strong enough to catch animals such as lizards, as well as insects. This spider is considered one of the most dangerous in the world. Its venom is strong enough to kill a person, though since the development of antivenoms very few human beings have been badly hurt.

Male redbacks do not make their own webs but may sit on the edge of the female's web. Sometimes female redbacks eat the male after mating.

The female

## ANT-MUGGING FLY

has a special trick for stealing food from a certain kind of ant. The fly approaches the ant and grabs hold of its antennae with its own. The fly then taps the ant, causing it to regurgitate its food. The fly lets go of the ant and steals the meal.

# The SPOOR SPIDER

lives in the baking hot Namib Desert in Africa. It spins its mat-like web of sticky silk on the sandy ground and then waits below it in a silk-lined burrow. When ants wander nearby, the spider springs out and presses them to the baking hot desert sands, cooking them alive!

# ATTACK AND DEFENCE

## Chapter 6

# BULLET ANTS

have one of the most painful stings of any insect. The pain the sting causes has been compared to being hit by a bullet – that's how this ant got its common name.

Worker bullet ants are large, up to 2.5 centimetres long, and they use their stings when hunting and to defend themselves. They live in Central and South American rainforests.

# BOMBARDIER BEETLES

are only up to a centimetre long, but they have a very effective and unusual way of defending themselves .

The beetle stores 2 kinds of chemicals in its body. If it is attacked, these chemicals are released into a special area of the beetle's abdomen where they mix to create an explosive substance. This boiling hot mixture reaches about 100 degrees centigrade, and is sprayed out of the beetle's anus in pulses at the enemy. Any sensible attacker will make a hasty retreat.

# STICK INSECTS

seem like peaceful creatures as they sit munching leaves. But there is a species that has a nifty defence if attacked. It can squirt out a thick smelly fluid to a distance of up to 40 centimetres to warn off predators such as mice and birds.

Human victims have reported suffering stinging pains in the eyes.

# KILLER BEES

are much more aggressive than other honeybees. They are quick to attack anything that disturbs them and they chase their enemies in greater numbers than other bees and for longer distances.

Stay well away from

# JACK JUMPER ANTS!

These ants live in Australia. If their colony is disturbed they can be extremely aggressive. They jump and attack anything they see as a threat and their sting contains powerful venom.

People have died from allergic reactions to the stings of these ants.

# FANTASTIC FEATS

## Chapter 7

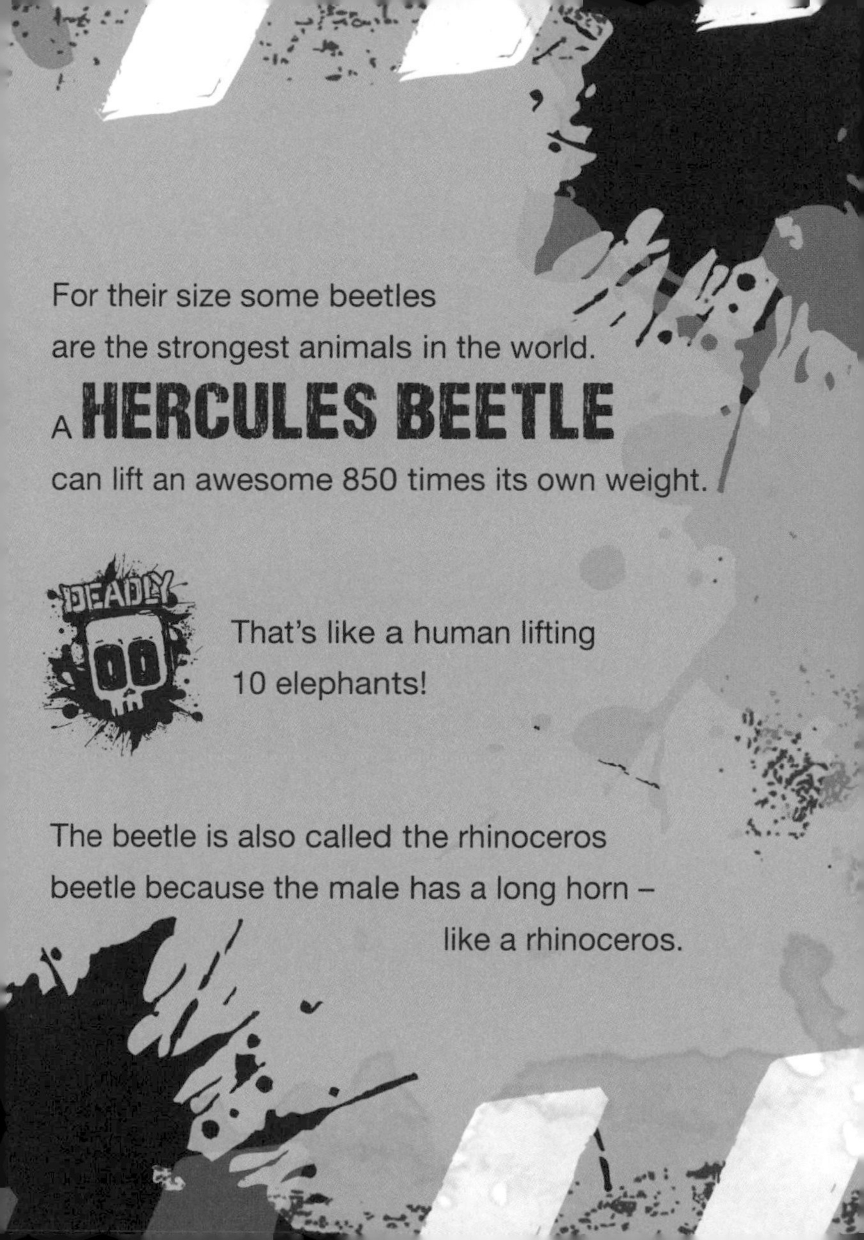

For their size some beetles
are the strongest animals in the world.

A **HERCULES BEETLE**
can lift an awesome 850 times its own weight.



That's like a human lifting
10 elephants!

The beetle is also called the rhinoceros
beetle because the male has a long horn –
like a rhinoceros.

# Male EMPEROR MOTHS

have an impressive sense of smell. Using the sensors on their large feathery antennae, some kinds have been known to sniff out females nearly 11 kilometres away.

SMALL EMPEROR MOTH

Can you believe that there are spiders living high on Mount Everest? The **HIMALAYAN JUMPING SPIDER** manages to survive at 6,700 metres above sea level. Its scientific name means 'standing above all others' and it is believed to feed on tiny insects that have been blown up the mountain slopes by the wind.

Male **CICADAS** make the loudest sounds of any insect as they 'sing' to attract females. The African cicada's call reaches a sound level of 107 decibels, louder than a hand-drill. The cicada's noisy call has a useful side-effect too – it actually puts off birds which might otherwise prey on the insects.

The cicada makes the sound by vibrating structures called tymbals on its abdomen with special muscles.

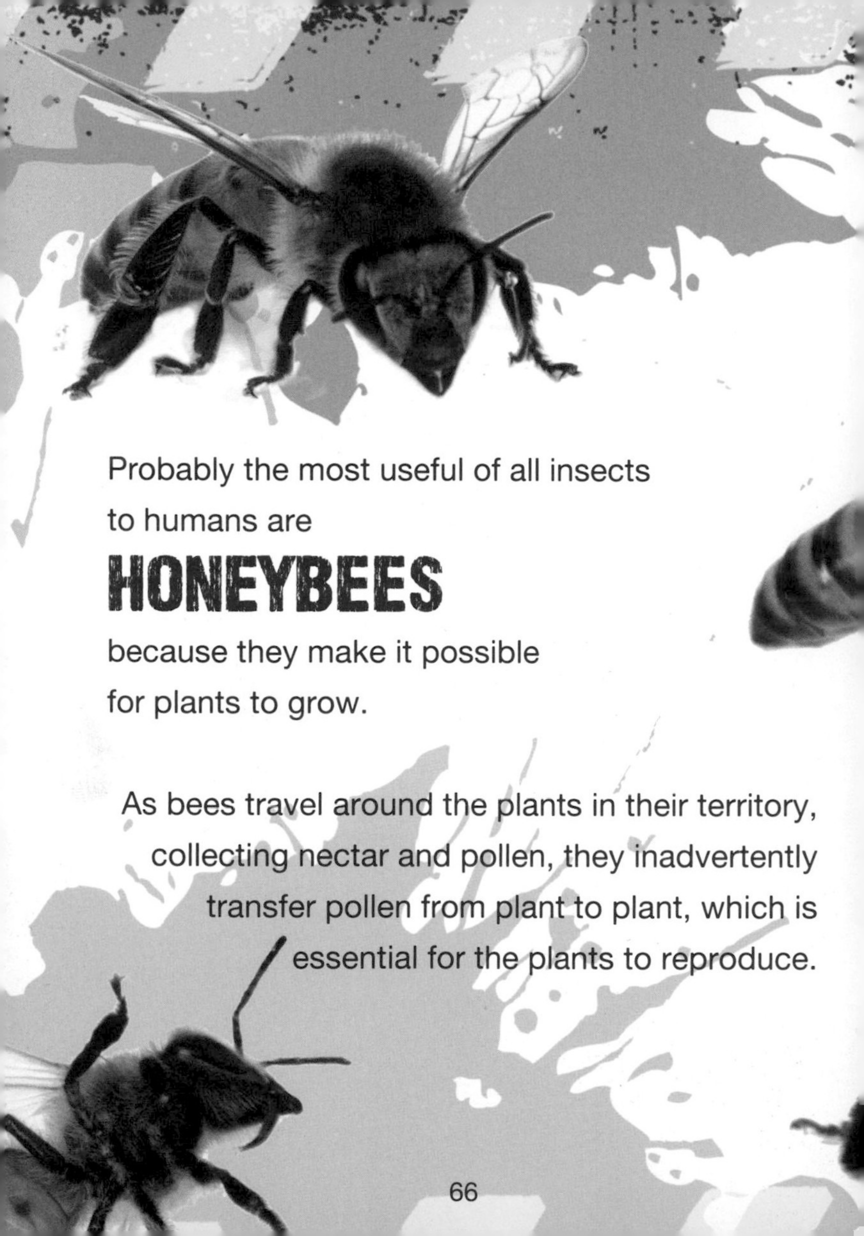

Probably the most useful of all insects to humans are

# HONEYBEES

because they make it possible for plants to grow.

As bees travel around the plants in their territory, collecting nectar and pollen, they inadvertently transfer pollen from plant to plant, which is essential for the plants to reproduce.

Bees also make honey as
food reserves for the winter.
It is said that bees must
make about 10 million
nectar-collecting trips to produce
enough honey to fill a 450 gram jar.

# The RAFT SPIDER

is the largest spider in Britain. It spends much of its life half in and half out of the water and can even skate over the water surface.

RAFT SPIDER ON WATER SURFACE

But it can do something even more incredible – it can stay under the surface for as long as an hour. It does this by using air bubbles trapped under the hairs on its body.

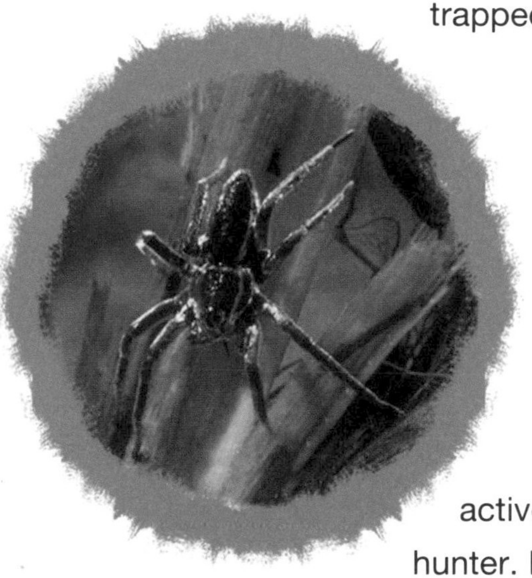

This spider is an active and aggressive hunter. It skates over the surface to catch insects, and dives beneath the water to ambush tadpoles, water boatmen and even small fish. Once it has them in its grasp, it injects paralysing venom which also turns the prey animal into easily digestible mush.

**FLEAS** are champion jumpers. These tiny insects can jump 100 times their own body length – imagine a person jumping to the top of a 40 storey building!

Fleas live by sucking the blood of other animals, such as cats and dogs. They need to leap high and fast in order to jump up on to their hosts – the animals they live and feed on.

# FAVOURITE FOODS

## Chapter 8

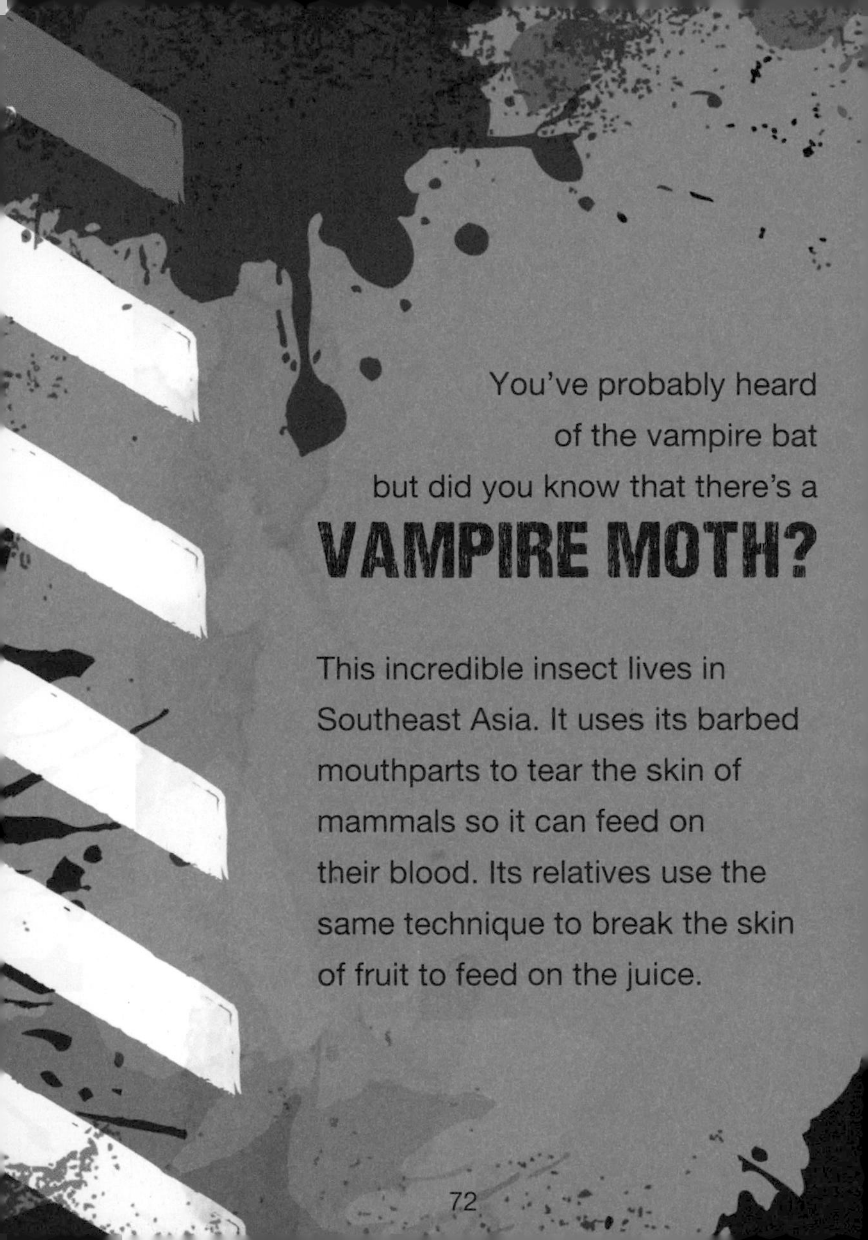

You've probably heard
of the vampire bat
but did you know that there's a

# VAMPIRE MOTH?

This incredible insect lives in Southeast Asia. It uses its barbed mouthparts to tear the skin of mammals so it can feed on their blood. Its relatives use the same technique to break the skin of fruit to feed on the juice.

A meal of tears has to be one of the strangest ever. Scientists have known for a while that some moths drink the tears of some animals such as deer and horses. Now they've discovered a moth in Southeast Asia that actually irritates the eyes of its victim in order to produce tears.

The moth inserts its mouthparts under the eyelids while the animal sleeps and brushes them across the eyeball to make tears flow. The tears may provide the moth with salt as well as fluid.

Another moth sucks fluids from the sores on an elephant's skin!

# American COCKROACHES

will eat more or less anything. Rotting plants and animal carcasses are their usual food, but they will also eat paper, book bindings, cloth, dead insects and hair.

There are about 3,500 species of cockroach living all over the world and they are among the most common of all insects.

Nectar is the main food of male and female **MOSQUITOES**.

But females also feed on the blood of other animals, including humans, in order to get enough protein to produce their eggs.

The female has special tube-like mouthparts. She punctures the victim's skin and injects special saliva that helps to keep the blood flowing as she sucks it up.

Because they can transmit deadly diseases as they bite, mosquitoes are the most dangerous of all animals. They cause large numbers of human deaths around the world.

# TARANTULA HAWK WASPS

are the largest wasps in the world. These deadly hunters prey on tarantulas, catching them to provide food for their young.

The female wasp attacks spiders on the ground and paralyses her catch with her sting. She then drags the helpless spider to a burrow, lays her egg on it and leaves. When the young wasp hatches, it feeds on the spider, which eventually dies. The adult wasp feeds on flower nectar.

# Look out for DEADLY DVDs, games and toys!

**The first Deadly DV**
Steve Backshall counts do
the most dangerous anim
in the world and picks
his personal top t
Release da
**3rd September 20**
**www.bbcshop.c**

Find your wild side with the new range of Deadly 60 games available from **www.thegameshop.co.uk**

Build your own Deadly 60 adventures with Character Building micro-figures and playsets. Find out more at **www.character-online.com**